ISC SECURITY

DESIGN CRITERIA

FOR NEW FEDERAL OFFICE BUILDINGS AND MAJOR MODERNIZATION PROJECTS

A REVIEW AND COMMENTARY

Committee to Review the Security Design Criteria
of the Interagency Security Committee

Board on Infrastructure and the Constructed Environment

Division on Engineering and Physical Sciences

NATIONAL RESEARCH COUNCIL
OF THE NATIONAL ACADEMIES

THE NATIONAL ACADEMIES PRESS
WASHINGTON, D.C.
www.nap.edu

THE NATIONAL ACADEMIES PRESS 500 Fifth Street, N.W. Washington, DC 20001

NOTICE: The project that is the subject of this report was approved by the Governing Board of the National Research Council, whose members are drawn from the councils of the National Academy of Sciences, the National Academy of Engineering, and the Institute of Medicine. The members of the committee responsible for the report were chosen for their special competences and with regard for appropriate balance.

This study was supported by Contract No. 45640000 between the National Academy of Sciences and the General Services Administration and Contract No. SALMEC-01-M-0545 between the National Academy of Sciences and the U.S. Department of State. Any opinions, findings, conclusions, or recommendations expressed in this publication are those of the authors and do not necessarily reflect the views of the organizations or agencies that provided support for this project.

International Standard Book Number 0-309-0888-01

Available from:

Board on Infrastructure and the Constructed Environment
National Research Council
500 Fifth Street, N.W.
Washington, DC 20001

Printed in the United States of America

THE NATIONAL ACADEMIES

Advisers to the Nation on Science, Engineering, and Medicine

The **National Academy of Sciences** is a private, nonprofit, self-perpetuating society of distinguished scholars engaged in scientific and engineering research, dedicated to the furtherance of science and technology and to their use for the general welfare. Upon the authority of the charter granted to it by the Congress in 1863, the Academy has a mandate that requires it to advise the federal government on scientific and technical matters. Dr. Bruce M. Alberts is president of the National Academy of Sciences.

The **National Academy of Engineering** was established in 1964, under the charter of the National Academy of Sciences, as a parallel organization of outstanding engineers. It is autonomous in its administration and in the selection of its members, sharing with the National Academy of Sciences the responsibility for advising the federal government. The National Academy of Engineering also sponsors engineering programs aimed at meeting national needs, encourages education and research, and recognizes the superior achievements of engineers. Dr. Wm. A. Wulf is president of the National Academy of Engineering.

The **Institute of Medicine** was established in 1970 by the National Academy of Sciences to secure the services of eminent members of appropriate professions in the examination of policy matters pertaining to the health of the public. The Institute acts under the responsibility given to the National Academy of Sciences by its congressional charter to be an adviser to the federal government and, upon its own initiative, to identify issues of medical care, research, and education. Dr. Harvey V. Fineberg is president of the Institute of Medicine.

The **National Research Council** was organized by the National Academy of Sciences in 1916 to associate the broad community of science and technology with the Academy's purposes of furthering knowledge and advising the federal government. Functioning in accordance with general policies determined by the Academy, the Council has become the principal operating agency of both the National Academy of Sciences and the National Academy of Engineering in providing services to the government, the public, and the scientific and engineering communities. The Council is administered jointly by both Academies and the Institute of Medicine. Dr. Bruce M. Alberts and Dr. Wm. A. Wulf are chair and vice chair, respectively, of the National Research Council.

www.national-academies.org

COMMITTEE TO REVIEW THE SECURITY DESIGN CRITERIA OF THE INTERAGENCY SECURITY COMMITTEE

STUART L. KNOOP, *Chair*, Oudens and Knoop Architects PC, Chevy Chase, Maryland
ALFREDO H.-S. ANG, University of California, Irvine
NIALL KELLY, Langdon Wilson Architects, Los Angeles, California
BRIAN MEACHAM, Ove Arup and Partners, Westborough, Massachusetts
RANDALL NASON, C.H. Guernsey & Company, Oklahoma City, Oklahoma
CHARLES OSWALD, Wilfred Baker Engineering, Inc., San Antonio, Texas
THOMAS RUST, Patton Harris Rust & Associates, PC, Chantilly, Virginia
KENNETH SCHOONOVER, KMS Associates, Inc., Lansing, Illinois
ROBERT SMILOWITZ, Weidlinger Associates, New York, New York
JAMES C. SNYDER, University of Michigan, Ann Arbor
LEONARD C. ZIMMERMANN, Flack & Kurtz, Inc., New York, New York

Staff

RICHARD G. LITTLE, Director, Board on Infrastructure and the Constructed Environment
JASON DREISBACH, Research Associate
DANA CAINES, Financial Associate
PAT WILLIAMS, Senior Project Assistant

BOARD ON INFRASTRUCTURE AND THE CONSTRUCTED ENVIRONMENT

Acknowledgment of Reviewers

This report has been reviewed in draft form by individuals chosen for their diverse perspectives and technical expertise, in accordance with procedures approved by the National Research Council's Report Review Committee. The purpose of this independent review is to provide candid and critical comments that will assist the institution in making its published report as sound as possible and to ensure that the report meets institutional standards for objectivity, evidence, and responsiveness to the study charge. The review comments and draft manuscript remain confidential to protect the integrity of the deliberative process. We wish to thank the following individuals for their review of this report:

John Crawford, Karagozian & Case
Kent Goering, Applied Research Associates
Eve Hinman, Hinman Consulting Engineers, Inc.
Douglas Mitten, Project Management Services, Inc.
Harold O. Sprague, Jr., Black & Veatch Special Projects Corp.

Although the reviewers listed above have provided many constructive comments and suggestions, they were not asked to endorse the conclusions or recommendations, nor did they see the final draft of the report before its release. The review of this report was overseen by Harold Forsen, National Academy of Engineering. Appointed by the National Research Council, he was responsible for making certain that an independent examination of this report was carried out in accordance with institutional procedures and that all review comments were carefully considered. Responsibility for the final content of this report rests entirely with the authoring committee and the institution.

Contents

List of Figures and Tables

FIGURES

TABLES

Executive Summary

The bombings of the World Trade Center in 1993, the Murrah Building in 1995, Khobar Towers in 1996, and the U.S. embassies in Nairobi and Dar es Salaam in 1998 made it clear that there are no safe havens from terrorism. Tragedy can strike anyone, anywhere, at any time—as we learned all too painfully on September 11, 2001.

Although the terrorist attacks of September 11 were shocking in their ferocity and wanton disregard for human life, they are acts that will not be easily repeated. The vehicle bomb employing conventional explosives still appears to be the most serious bomb threat confronting federal buildings. Given the large number of buildings potentially vulnerable to terrorist attack, it is imperative that their planning and design be sound, particularly when the magnitude of risk will vary widely depending on each facility's mission and location.

For both new facilities and major renovations, physical protection must be integrated into the planning and design process. The challenge for design teams is to create facilities that protect against terrorist explosive threats while at the same time retaining the features that make them desirable workspaces. Often situated on urban sites, federal buildings are limited in the options available to restrict access to them or to provide effective standoff distances. Architectural and site design aesthetics are also difficult to balance with blast mitigation.

After the bombing of the Alfred P. Murrah Federal Building in Oklahoma City, the U.S. Marshal's Service assessed the vulnerability of federal buildings. The General Services Administration (GSA), in concert with other federal agencies, was then directed by Executive Order 12977 to set construction standards for buildings that required specialized security measures, such as blast resistance. The result was the Interagency Security Committee (ISC) *ISC Security Design Criteria for New Federal Office Buildings and Major Modernization Projects* (ISC, 2001). The criteria apply to new construction or major renovation of those office buildings and courthouses occupied by federal employees in the United States that are not under the jurisdiction or control of the Department of Defense, which has its own criteria. The ISC criteria reflect a "flexible and realistic approach to the reliability, safety, and security of Federal office buildings."

In November 1999, GSA and the U.S. Department of State convened a symposium to discuss the apparently conflicting objectives of security from terrorist attack and the design of public buildings in an open society. The symposium sponsors rejected the notion of rigid, prescriptive design approaches. The symposium concluded with a challenge to the design and security professions to craft aesthetically appealing architectural solutions that achieve balanced, performance-based approaches to both openness and security.

In response to a request from the Office of the Chief Architect of the Public Buildings Service, the National Research Council (NRC) assembled a panel of independent experts, the Committee to Review the Security Design Criteria of the Interagency Security Committee. This committee was tasked to evaluate the ISC Security Design Criteria to determine whether particular provisions might be too prescriptive to allow a design professional "reasonable flexibility" in achieving desired security and physical protection objectives.

EVALUATION OF THE ISC SECURITY CRITERIA

The committee has performed a detailed, line-by-line evaluation of the criteria and commented at length to the GSA on certain provisions. The committee finds the *ISC Security Design Criteria* document in general to be a mix of performance objectives, prescriptive requirements, and references to industry design standards. The committee believes that although full implementation of all the ISC criteria will provide some protection for building occupants against most blast-related threats and should significantly reduce injuries, the organization of the document makes it difficult to identify clearly the connections between specific criteria and the performance objectives they are meant to achieve. It is also difficult to identify clearly how some criteria apply to specific components of building design. Because this is a critical element in a performance-based design process, the committee believes that rectifying this shortcoming should be given priority. Another concern is that because the document is focused on the terrorist vehicle bomb as the primary means of attack, it offers little guidance on defending federal buildings and their occupants from other terrorist actions, including those using chemical, biological, or radiological weapons.

The committee is troubled by what it perceives to be a disconnect between the authorizing action for the security criteria (Executive Order 12977) and application of the criteria in practice. Without a clear and explicit statement that mandates use of the ISC criteria for all covered projects, it appears that the criteria will be applied at the discretion of individual project managers in GSA and other federal agencies—a highly undesirable situation. Indeed, other agencies are already drafting their own criteria independent of the ISC.

These and related issues can be addressed by further refinement of the *ISC Security Design Criteria* document—clarifying and reformatting it to provide more complete and comprehensible design criteria as well as necessary policy guidance. The committee believes that the continued use of both prescriptive and performance-based criteria is appropriate for several reasons, including the fact that most buildings can be designed quite effectively and flexibly using prescriptive criteria. Performance analysis and design are needed only for selected portions of the process; structured appropriately, prescriptive criteria—augmented by good practices from lessons learned—can themselves become a means of meeting performance objectives.

The committee recommends that several short-term actions be initiated if the ISC is truly aiming for a more performance-based approach to security-related design. This is especially important if design is to be based on clear and explicit guidance and fully integrated with other aspects of the facility planning and design process.

The following recommendations address issues that the committee believes should receive the immediate attention of the ISC; unless these first recommendations are implemented promptly, the ISC Security Design Criteria will probably continue to be underused and often misinterpreted by users who do not have a strong background in blast or security analysis. Additional recommendations are presented in Chapter 4, but the committee does not view them as having similar urgency.

GROUP 1—SHORT-TERM RECOMMENDATIONS

The committee recommends that:

1. The Interagency Security Committee give immediate consideration to the changes suggested by the committee as a result of its evaluation of the ISC Security Design Criteria.

2. The *ISC Security Design Criteria* be expanded to include basic information on how blasts affect buildings and people, as a technical basis for the protective strategies incorporated in the document (for suggested content, see Chapter 3 below).
3. A preface be added to the *ISC Security Design Criteria* that clearly establishes the authority and applicability of the document by citing Executive Order 12977 and all related executive or legislative actions.
4. The *ISC Security Design Criteria* be reviewed by a professional editor familiar with building codes and standards to improve its organization and ensure consistency in format, language, and style. This should improve both the readability and the utilization of the document.
5. Information provided in Part 3 of the *ISC Security Design Criteria* document be integrated with Parts 1 and 2, so that all information pertaining to a particular design component or system (structural or mechanical) is located together, thus clarifying the document and simplifying its use.
6. The spirit and intent of the language in the introduction to the *ISC Security Design Criteria* be incorporated in any document that references the criteria.
7. The Interagency Security Committee in concert with its member agencies begin a comprehensive review of the ISC Security Design Criteria as soon as possible. This review should include creation of risk assessment and management tools as well as policy guidance for physical protection and security to guide the development of risk reduction strategies and a performance-based design process. The review should also consider expanding the criteria to (1) include critical occupancies such as child care centers, (2) broaden the range of possible threats and countermeasures to cover chemical, biological, and radiological weapons, and (3) address the impact on protective design of such issues as emergency planning, enhanced fire protection, life-cycle costing, procurement strategy, and construction security.

REFERENCE

ISC (Interagency Security Committee). 2001. ISC Security Design Criteria for New Federal Office Buildings and Major Modernization Projects. Washington, D.C.: General Services Administration.

1

Introduction

BACKGROUND

In the aftermath of the bombing of the Alfred P. Murrah Federal Building in Oklahoma City, the Interagency Security Committee (ISC) was established on October 19, 1995, by Executive Order 12977 (Clinton, 1995) and charged to set long-term construction standards for locations requiring blast resistance or other specialized security measures. The ISC drafted the *ISC Security Design Criteria for New Federal Office Buildings and Major Modernization Projects* (ISC, 2001) to ensure that security becomes an integral part of the planning, design, and construction of these projects. Both the ISC Security Design Criteria and the GSA Draft Security Design Criteria that preceded it grew out of the Department of Justice's Vulnerability Assessment (DOJ, 1995), written after the 1995 Oklahoma City bombing.

The criteria are intended to employ a "flexible and realistic approach to the reliability, safety and security of Federal office buildings" and recognize that they may not be achievable in all cases. In November 1999 the GSA and the U.S. Department of State convened a symposium to discuss the apparently conflicting objectives of providing security from terrorist attack while designing public buildings in an open society. The sponsors rejected the notion of rigid, prescriptive design approaches. The symposium concluded with a challenge to the design and security professions to find aesthetically appealing architectural solutions that achieve both security and physical protection; a balanced, performance-based approach to security and openness.

After the symposium, the Chief Architect of the GSA Public Buildings Service and the U.S. Department of State asked the NRC to evaluate the ISC Security Design Criteria and offer recommendations for improvement.

SCOPE OF THE REVIEW

In response to that request, the NRC assembled a panel of independent experts, the Committee to Review the Security Design Criteria of the Interagency Security Committee, under the auspices of the NRC Board on Infrastructure and the Constructed Environment. The 11 members of the committee have expertise in architecture, structural engineering, blast-effects mitigation, mechanical and electrical engineering, physical security, site design, risk analysis and management, facility planning and cost engineering, and performance-based building codes. Biographical information about the committee members is provided in the appendix.

The committee was charged with three specific tasks:

1. Review and evaluate the ISC Security Design Criteria to determine whether particular provisions could be considered prescriptive and not performance-based. Where criteria are judged by the committee to be too prescriptive to give a design professional reasonable flexibility in achieving the objective, performance-based approaches are to be suggested.
2. Discuss the integration of a performance-based design approach into the decision-making process identified in the ISC Security Design Criteria.
3. Identify practices, documents, and tools in addition to those listed in the ISC Security Design Criteria that are necessary to implement a performance-based system for security and blast-resistant design.

ORGANIZATION OF THE REPORT

While performing its first task, the committee recognized the desirability of objective criteria upon which to base its evaluation, but it became immediately apparent that no objective criteria were available. Therefore, the committee determined that the most reasonable approach would be to use the professional expertise and experience of its members to assess whether a specific provision or design criterion would offer a design professional "reasonable flexibility and creativity" in achieving the desired security and physical protection objectives. If it did not, the provision was considered overly prescriptive and alternative language proposed. If it did, this was noted but the committee often provided clarifying language. The results of the committee's line-by-line review of the ISC Security Design Criteria, made available to the sponsor and not included in this report, are presented in a two-column, commentary format with the language of the criteria in the left column and any comments from the committee or proposed changes to that language in the right column. Besides commenting on the specific provisions of the ISC Security Design Criteria, the committee also addressed the broader issue of performance-based design, as required by the second task with which it was charged.

Performance-based regulatory processes link specific design requirements for a building to broader social, political, or economic goals. For the most part, this kind of guidance is absent from the ISC Security Design Criteria. In Chapter 2 below the committee expands on the performance-based design process, describing the steps necessary for the ISC to design a meaningful performance-based process for design of security-related mitigation of blast effects.

During its review, the committee noted that the ISC Security Design Criteria do not contain enough background information on how blasts affect buildings and people to provide a context for a security design process that can be implemented without prescriptive requirements. Chapter 3 therefore proposes a primer on blast effects on buildings and people for users of the ISC criteria that describes the nature of explosives, the effects of explosions on buildings and people, and design strategies to improve protection. This material is written primarily for the reader without specialized experience or training in blast design. It includes information that will be useful to officials from GSA and other federal agencies and their architects, engineers, and tenants (among others). This chapter addresses the third task with which the committee was charged.

The results of the committee's assessment and its recommendations are presented in Chapter 4. The recommendations are presented in two groups. The first are those the committee believes can and should be implemented in the short term to help maximize comprehension and usage of the important security concepts in the criteria with minimum effort. The second group addresses fundamental ways the ISC Security Design Criteria could be improved in the longer term through both amendments and improvements to government oversight of their application.

REFERENCES

Clinton, William Jefferson. 1995. Executive Order 12977, Interagency Security Committee. Washington, D.C.: The White House.

DOJ (Department of Justice). 1995. Vulnerability Assessment of Federal Facilities. Washington, D.C.: U.S. Department of Justice.

ISC (Interagency Security Committee). 2001. ISC Security Design Criteria for New Federal Office Buildings and Major Modernization Projects. Washington, D.C.: General Services Administration.

2
A Performance-based Approach to Security and Blast-mitigating Building Design

INTRODUCTION

One of the tacit assumptions underlying the committee's charge was that the current ISC Security Design Criteria (ISC, 2001) do not give a designer enough flexibility to explore alternative (presumably more creative) approaches to security. Without rendering judgment on that assumption, in this chapter the committee describes a hypothetical performance-based approach for designing buildings subject to the ISC criteria that is modeled on the *International Performance Code for Buildings and Facilities* (ICC, 2001). This material is not presented to advocate a specific approach but rather to illustrate that setting up a formal performance-based system is not a simple matter and instead requires considerable forethought.

Although the committee believes that the material presented here could serve as a model for a performance-based security design process, the wording of the various process elements is for illustration only; it *should not* be considered a specific recommendation of the committee for the ISC Security Design Criteria. It is the committee's belief that setting specific goals and criteria for performance-based security design of federal buildings is an inherently governmental function, to be undertaken by the ISC itself, perhaps assisted by experts in drafting building codes and engineering standards. If the ISC wishes to implement a more formal performance-based approach for security-related design, crafting the necessary policy guidance and performance criteria should be given a high priority, as should development of the other process elements.

A POSSIBLE PERFORMANCE-BASED FRAMEWORK FOR THE ISC SECURITY DESIGN CRITERIA

Background

"Prescriptive" or specification-based systems describe how buildings should be designed, built, protected, and maintained. For the most part, this is done through codes and standards that prescribe (specify) *what* is required for health, safety, and general welfare, *how* these requirements are to be met, and how compliance is to be *verified*.

A performance-based system, on the other hand, does not limit the designer to prescribed construction details or design requirements. Rather, it relates the design and construction of the building to the desired performance of the building under certain conditions (in the case of the ISC Security Design Criteria, a terrorist attack using conventional explosives) based on preset performance objectives. A performance-based system provides a policy framework for defining the building performance desired and permits the use of a variety of acceptable methods to meet the requirements.

Performance-based commercial building codes typically have three components (Meacham, 1998a):

1. *Codes (regulations)* that establish the expected level of performance through explicit statements of *goals*, *functional objectives,* and *performance requirements*;
2. *Engineering standards, guides, and practices*, which are separate documents, adopted in the codes by reference, that describe *acceptable methods* for complying with a code; and
3. *Evaluation and design tools*, *acceptable methods* for creating, reviewing, and *verifying* designs in accordance with *engineering standards, guides,* and *practices.*

Acceptable methods are design approaches, material testing procedures, or prescriptive requirements that may be included in or referenced by a code. They may be specified (prescribed) solutions or standards, practices, tools, and methodologies that can be used for both design and verification of compliance.

In a performance-based system, there is a clear differentiation between:

1. The requirements of the code or regulation (the *what* component),
2. Acceptable means for complying with the code (the *how* component), and
3. Acceptable means for demonstrating that the proposed solutions comply with the code (the *verification* component).

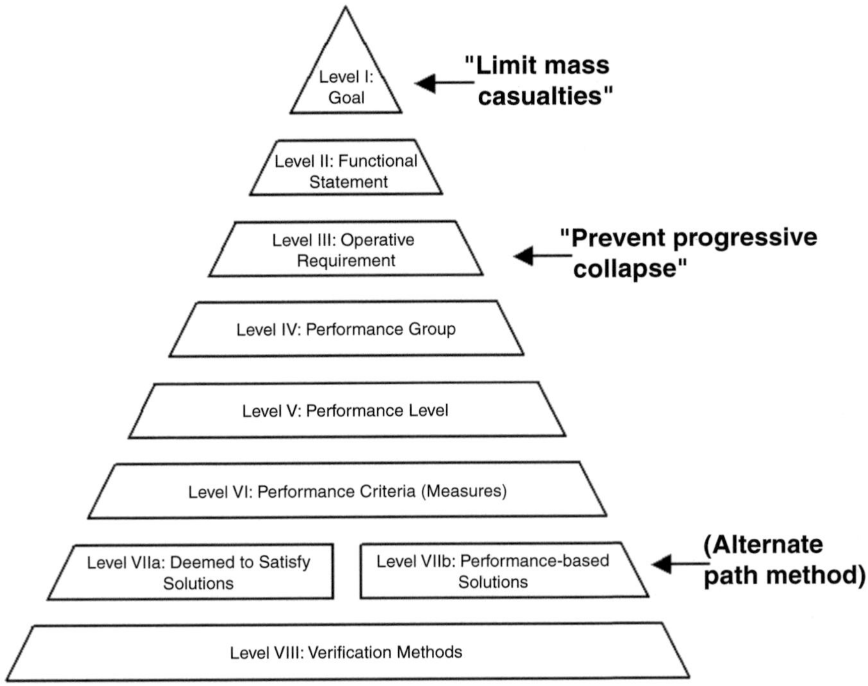

FIGURE 2.1 Performance-based building code hierarchy.

For performance-based regulations and design methods to be effective, there must be a logical and transparent relationship between what the regulations are designed to achieve and the methods used to achieve it. An eight-level hierarchy (see Figure 2.1) displays these relationships. This hierarchy illustrates how the goal "limit mass casualties" can be made operative through a performance requirement to "prevent progressive collapse" and achieved through a design approach (the alternate path method) that can be demonstrated to be effective. The important role played by test methods and standards, evaluation methods, design guides, and other verification methods is also readily seen.

A Policy Framework for Security Design

Goals

The highest level of the building performance hierarchy consists of policy or societal goals. In general, goals for the built environment should reflect the interests both of the community at large and of specific stakeholders, such as owners, tenants, and facility managers. Goals for commercial building codes are often limited to health, safety, welfare, amenity, and accessibility, and sometimes property protection, affordability, and durability. Additional goals may be needed, such as mission continuity. To be effective, goals should be clear, unambiguous, and easily comprehensible; they should not be described in technical terms. Policy goals are most appropriately established by governmental or quasi-governmental bodies that understand the building construction and regulatory issues and that are responsible for the health and safety of those who occupy the regulated buildings. In many regulatory systems, policy goals ultimately become part of the laws regulating buildings.

For the ISC Security Design Criteria, a section on policy goals might include statements like the following (adapted from ICC, 2001):

- The purpose of these criteria is to help design facilities that save lives, prevent injuries, and protect critical functions and assets from a bombing or other physical attack.
- The primary goal for facilities constructed using these guidelines is to protect the occupants from death and injury and to limit damage to the facility, its contents, and its mission from acts of terrorism and related violence that may reasonably be expected. Full implementation of these guidelines should provide some protection against all threats and significantly reduce casualties. These guidelines therefore intend that federal facilities provide for:

    1. Minimizing the risk of death and injury from natural and technological hazards, acts of terrorism, and acts of social violence.
    2. A structure that will reasonably withstand those loads that can be expected to impact the facility throughout its intended lifetime.
    3. Means of egress and access for emergency as well as normal circumstances.
    4. Methods of limiting the spread of fire, products of combustion, and airborne contaminants both within the building and to adjacent properties.

Once clear, unambiguous, and easily comprehensible goal statements have been developed, more detailed functional and operative (performance) statements can be established.

Functional Statements

For each goal, functional statements set the qualitative requirements for buildings or specific building elements (features), describing how the goal can be met. To meet the goal for "minimizing risk of injury and death," for instance, functional statements might deal with such factors as blast protection, fire safety, structural stability, and protection from intrusion. Because functional statements provide more specific guidance than the general goal statement, they should contain or reference a measure (qualitative or quantitative) of the level of performance that is tolerable or acceptable to the community of stakeholders. This may be done by setting specific criteria, referring to "deemed to satisfy" solutions, or referring to criteria, methods, or standards published by others.

The following example illustrates how functional statements might be formulated:

- The primary goal for facilities constructed using these guidelines is to protect the occupants from death and injury and to limit damage to the facility, its contents, and its mission from acts of terrorism and related violence that may reasonably be expected to occur.

This statement of goals speaks to the issues of safety, damage, and mission continuity but gives no specific guidance on functional expectations. Associated functional statements are needed. The following are examples of functional statements:

- **Life safety and injury prevention.** Structures shall be designed and constructed so as to prevent injury to occupants due to the failure of a structural element or system loaded to levels determined by a facility-specific risk assessment.
- **Property and amenity protection.** Structures shall be designed and constructed to prevent loss of property and amenities due to the failure of a structural element or system loaded to levels determined by a facility-specific risk assessment.
- **Mission continuity.** Structures shall be designed and constructed to prevent loss of mission continuity due to the failure of a structural element or system loaded to levels determined by a facility-specific risk assessment.

Performance Requirements

Operative or performance requirements state actual requirements in terms of performance criteria or expanded functional descriptions. Like functional statements, they should refer to acceptable or tolerable levels of performance. There is considerable flexibility in how this might be accomplished. Some operative requirements may simply be detailed qualitative statements; others might specify quantitative performance criteria. The choice may depend on the degree of flexibility desired, or may be limited by the number of solutions available. The following are examples of qualitative performance statements for selected performance categories:

- **Stability.** Structures or portions thereof shall remain stable and not collapse during construction or alteration and throughout their lives.
- **Disproportionate failure.** Structures shall be designed to sustain local damage, and the structural system as a whole shall remain stable and not be damaged to an extent disproportionate to the original local damage.
- **Progressive failure.** Structures shall be designed to prevent progressive collapse due to the loss of one primary column, or it shall be demonstrable that the proposed design precludes such a failure.
- **Exterior walls.** Exterior walls shall be designed to resist the actual pressures and impulses acting on exterior wall surfaces from the threats defined for the facility. Exterior walls shall be capable of withstanding the dynamic reactions from the windows.
- **Loss of amenity.** Structures or portions thereof shall have a low probability of causing damage or loss of amenity through excessive deformation, vibration, or degradation during construction or alteration and throughout their lives.

The following example illustrates how quantitative performance criteria can be integrated into operative statements.

- **Exterior walls:** To comply with Medium Protection Levels, exterior walls should be designed to resist elastically a pressure of X psi and an impulse loading of Y psi-msec; for Higher Protection Levels, pressure to be resisted is X′ psi and impulse loading is Y′ psi-msec.

Performance Groups

In performance-based regulatory systems, building-use or occupancy types for which similar performance is desired or required are often arranged into performance groups (ICC, 2001; Meacham, 1998a). The primary value to the ISC of such groupings would be to simplify security planning and design for similar types of facilities or facilities subject to similar threats. For example, the Department of Justice Vulnerability Assessment (DOJ, 1995) grouped facilities by facility type, tenant, occupancy, and perceived risk (see Table 2.1). As a starting point, the ISC Security Design Criteria could use the DOJ levels as the basis for establishing such groupings.

TABLE 2.1 Security Levels for Federal Facilities

Level	Typical Location	Typical Tenants
I	≤10 Federal Employees ≤ 2,500 square feet Low-volume public contact Small "store-front" operation	Military Recruiting Small Post Office USDA Office Border Patrol Station (remote) Customs/INS Checkpoint (remote) Social Security Administration
II	11-150 Federal Employees 2500-80,000 square feet Moderate-volume public contact Routine operations similar to private sector or facility shared with private sector	Public Officials (Congress/Senate) Railroad Retirement Board INS Offices U.S. Customs Offices
III	151-450 Federal Employees Multistory facility 80,000-150,000 square feet Moderate/high-volume public contact Agency mix: Law enforcement operations Court functions Government records	U.S. Bankruptcy Court Inspector General IRS Criminal Investigations Division U.S. Probation Service U.S. Pretrial Services Federal Public Defender GSA Field Office
IV	> 450 Federal Employees Multistory facility > 150,000 square feet High-volume public contact High-risk law enforcement/ intelligence agencies District court	U.S. District Courts U.S. Marshals Service FBI DEA ATF U.S. Secret Service
V	Level IV profile and agency/mission critical to national security	U.S. Department of State HQ CIA Headquarters Pentagon White House U.S. Capitol

SOURCE: DOJ (1995).

Performance Levels

Performance levels describe the desired, required, or expected performance of a building or structure in terms of a specified measure. Establishing these levels requires balancing of technical considerations and societal values. The terms "tolerable" or "acceptable" are often used to reflect the fact that absolute protection is not possible and that some damage, injury, or loss will occur in structures, especially after a hazard event. The term "impact" is used as a broad descriptor of loss.

The ISC Security Design Criteria already embody this concept to some degree through the four protection levels (low, medium/low, medium, and higher). Part 3 of the criteria, the Risk Guidelines, specifies the desired level of performance (i.e., how the building is to perform during an emergency and the degree to which the building and its constituent elements should protect against specific tactics). The protection levels can be translated into performance terms that are definitive statements of what is desired or expected from the building for each level of protection and could read as follows.

Low Impact. The tolerable impacts of the design loads are as follows:

- **Structural damage.** There is no structural damage and the building or facility is safe to occupy.
- **Nonstructural systems.** Nonstructural systems needed for normal building or facility use and emergency operations are fully operational.
- **Occupant hazards.** Injuries to building or facility occupants are minimal in number and minor in nature. There is a very low likelihood of single or multiple life loss.*
- **Extent of damage.** Damage to building or facility contents is minimal in extent and minor in cost.*
- **Hazardous materials.** Minimal amounts of hazardous materials are released to the environment.

Low/Moderate Impact. The tolerable impacts of the design loads are as follows:

- **Structural damage.** There is moderate structural damage, which is repairable; some delay in reoccupancy can be expected.
- **Nonstructural systems.** Nonstructural systems needed for normal building or facility use are fully operational, although some cleanup and repair may be needed. Emergency systems remain fully operational.
- **Occupant hazards.** Injuries to building or facility occupants may be locally significant but generally moderate in number and in

*Applies only to hazard-related applied loads. Depending on the applied load (e.g., fire hazard) expected injuries and damage may be higher in localized areas, while the remaining areas sustain fewer injuries and less damage.

nature. There is a low likelihood of single life loss, very low likelihood of multiple life loss.*

- **Extent of damage.** Damage to building or facility contents, though it may be locally significant, is generally moderate in extent and cost.*
- **Hazardous materials.** Some hazardous materials are released, but the risk to the community is minimal. No emergency relocation is necessary.

Moderate Impact. The tolerable impacts of the design loads are as follows:

- **Structural damage.** There is significant damage to structural elements but no large quantity of falling debris; repair is possible. Significant delays in reoccupancy can be expected.
- **Nonstructural systems.** Nonstructural systems needed for normal building or facility use are significantly damaged and inoperable; egress routes may be impaired by light debris; emergency systems may be significantly damaged but remain operational.
- **Occupant hazards.** Injuries to building or facility occupants may be locally significant, with a high risk to life, but are generally moderate in number and nature. There is a moderate likelihood of single life loss, a low probability of multiple life loss.*
- **Extent of damage.** Damage to building or facility contents may be locally total and generally significant.*
- **Hazardous materials.** Hazardous materials are released, with localized relocation needed for occupants of buildings and facilities in the immediate vicinity.

Higher Impact. The tolerable impacts of the design loads are as follows:

- **Structural damage.** There is substantial structural damage but all significant components continue to carry gravity load demands. Repair may not be technically possible. The building or facility is not safe for reoccupancy, because reoccupancy could cause collapse.
- **Nonstructural systems.** Nonstructural systems for normal building or facility use may be completely nonfunctional. Egress routes may be impaired; emergency systems may be substantially damaged and nonfunctional.

---

*Applies only to hazard-related applied loads. Depending on the applied load (e.g., fire hazard) expected injuries and damage may be higher in localized areas, while the remaining areas sustain fewer injuries and less damage.

- **Occupant hazards.** Injuries to building or facility occupants may be high in number and significant in nature. Risk to life may be significant. There is a high likelihood of single life loss and a moderate likelihood of multiple life loss.*
- **Extent of damage.** Damage to building or facility contents may be total.*
- **Hazardous materials.** Significant amounts of hazardous materials are released, with relocation needed beyond the immediate vicinity.

Performance Criteria

Performance criteria are the metrics against which performance should be measured, predicted, and evaluated for compliance with goals, functional objectives, and performance requirements; these criteria are integral to a performance-based system.

Commercial building codes deal with structural loads, natural hazards, and technological hazards (e.g., fire, explosion, contamination, infection, and other safety and health hazards). As long as protection from these hazards achieves some tolerable level of performance and no other goals and objectives are violated, the intent of the codes is met. In a prescriptive code, the criteria for tolerable levels of performance are stated in a variety of forms (e.g., applied loads, air changes per hour, maximum travel distances). Unfortunately, the prescribed criteria often reflect what happens when formulae or standardized tests are idealized; they may not reflect actual performance under load, especially under the extreme conditions experienced during hazardous events.

In a performance-based regulatory system, there must be ways to define, assess, design, and evaluate the actual performance of building components, systems, and assemblies and of the structure itself. Thus, performance criteria (measures) are needed to serve as metrics for determining compliance. The code may state such criteria qualitatively as well as quantitatively, but quantitative values are required for design and evaluation.

However, determining tolerable levels of performance is not easy. In the case of safety, for example, it is difficult to determine the "load" against which appropriate "resistance" is required. This is certainly the situation with prescriptive fire protection requirements. Fire tests of building materials, elements, and systems generally use different, and sometimes incompatible or unrealistic, loads (Snell, 2001). When these materials and sys-

*Applies only to hazard-related applied loads. Depending on the applied load (e.g., fire hazard) expected injuries and damage may be higher in localized areas, while the remaining areas sustain fewer injuries and less damage.

tems are assembled in a building, it is difficult to state accurately how much performance (safety) is actually being provided.

To address these concerns, the *International Performance Code for Buildings and Facilities* uses the concepts of "design load" and "performance levels" (Meacham, 1998a; ICC, 2001). A "design load" is used because, although the maximum load a structure will be subjected to (especially hazard-induced loads) cannot be predicted, the structure can be designed to resist loads that can reasonably be expected. In the *International Performance Code,* design loads are divided into four categories of increasing magnitude: small, medium, large, and very large. "Small" loads indicate high-probability, limited-consequence events; "very large" loads indicate low-probability, significant-consequence events.

Definition of design loads also depends on how performance is to be measured. In the current ICC approach, performance is expressed in terms of tolerable impacts (consequences) on structures, occupants, and contents (ICC, 2001). Thus, the impact of a fire of a certain size in an office may be different from the impact of a similar event in a laboratory. As a result, the definition of a "small" design fire will differ for different types of facilities. Explicit statements of tolerable impacts to which the structure must conform when subjected to design loads of various magnitudes are provided through the performance levels.

The ISC criteria use a similar approach, but inconsistencies in their terminology could lead to confusion. For example, the "loads" are the "tactics" defined in Part 3 as low, medium/low, medium, and higher—the same classification system as for the protection levels, but here the terms refer to risk rather than performance. In Part 2, Table 4.1, "loads" refers to hazard levels, in the same taxonomy of low, medium/low, medium, and higher. To make the criteria more usable, the committee believes, the ISC should employ consistent terminology. It would also be clearer to have a different taxonomy for "loads" (tactics/hazards) and for "resistances" (protection levels). These could be pairings of small/frequent, medium/less frequent, large/rare, and very large/very rare.

A matrix (Table 2.2) can be constructed of performance groups, performance levels (protection levels), and performance criteria (tactics) to give criteria users a simple visual representation of the damage to be expected for different magnitudes of anticipated threats and for the four ISC protection levels.

"Deemed to Satisfy" Solutions

One way to demonstrate compliance with a performance-based regulation is to apply strictly prescriptive codes and standards (sometimes called "approved documents" or "approved methods") that have been "deemed

TABLE 2.2 Damage to Be Expected Based on Protection Levels and Design Event Magnitudes

		INCREASING LEVEL OF RISK →→→→→→→→→→→→→→→→→→→ PERFORMANCE GROUPS			
		Performance Group I (Protection Level Low)	Performance Group II (Protection Level Medium/Low)	Performance Group III (Protection Level Medium)	Performance Group IV (Protection Level High)
MAGNITUDE OF DESIGN EVENT (TACTIC) INCREASING MAGNITUDE OF EVENT ↑↑↑↑↑↑↑↑↑↑↑↑↑↑	VERY LARGE (Very Rare) (High Risk)	SEVERE	SEVERE	HIGH	MODERATE
	LARGE (Rare) (Medium Risk)	SEVERE	HIGH	MODERATE	MILD
	MEDIUM (Less Frequent) (Med/Low Risk)	HIGH	MODERATE	MILD	MILD
	SMALL (Frequent) (Low Risk)	MODERATE	MILD	MILD	MILD

SOURCE: Adapted from ICC (2001).

to satisfy" the goals, functional statements, and performance requirements of the regulation. "Deemed to satisfy" solutions may be used for technical, political, legal, or practical reasons (e.g., ease of use and enforcement or proof that an agreed "standard of care" has been exercised).

Often, "deemed to satisfy" solutions are derived from the prescriptive requirements that were in place before a performance-based code was introduced. Although this approach provides some performance criteria and a built-in comfort level, there is the chance that the prescriptive requirements (and associated criteria) do not, in fact, comply with the goals, performance objectives, and other elements of the performance-based system, due, perhaps, to new requirements or because metrics for assessing performance are unavailable or incompatible. This is why it is essential to link goals, functional statements, performance objectives, and criteria.

"Approved documents" to support "deemed to satisfy" solutions may be installation standards, prescriptive design standards (guidelines, codes of practice), and prescriptive codes (building, fire, plumbing, electrical, and mechanical). Most performance-based regulations allow for a mix of prescriptive and performance-based solutions because (1) performance-based solutions may be needed not for the entire building but only for specific components, (2) prescriptive solutions can be easier to apply, and (3) the combination of prescriptive and performance solutions can provide optimal flexibility and cost-effectiveness.

Performance-based Solutions

Performance-based solutions typically meet operative requirements and associated criteria by employing the analysis and design standards, guidelines, or practices generally accepted in the relevant professional disciplines. In the broadest sense, performance-based analysis and design is a process of designing a solution to meet specific performance goals that have been stated in terms of qualitative or quantitative objectives, criteria, or limitations of damage or injury (Meacham, 1998b). In structural engineering, for example, performance levels are often defined in terms of specific limiting damage states against which a structure's performance can be objectively measured (Hamburger et al., 1995). This approach has been adopted, for instance, by the fire safety engineering community, which defines performance-based fire safety design as:

> an engineering approach to fire protection design based on (1) agreed upon fire safety goals, loss objectives, and design objectives; (2) deterministic and probabilistic evaluation of fire initiation, growth, and development; (3) the physical and chemical properties of fire and fire effluents; and (4) quantitative assessment of the effectiveness of design alternatives against loss objectives and performance objectives. (Custer and Meacham, 1997)

The process of selecting performance-based analysis and design methods is outlined in Figures 2.2 and 2.3 (adapted from Meacham, 1998b). In Figure 2.2, the process is outlined in sequence; in reality, however, there will be several iterations, especially during the evaluation stage (Step 6). The iterative nature of the evaluation, including the possibility of revisiting selection of performance criteria, is illustrated in Figure 2.3.

The advantage of performance-based analysis and design lies in the flexibility that can be achieved without compromising safety, cost, or other important factors. It allows all parties to agree on goals, objectives, criteria, and analysis and evaluation methods, resulting in a design that best fits all parameters—a performance-based design solution.

Performance-based analysis and design can be used in conjunction with a performance-based regulation or on its own. The primary difference is that where a regulation already exists, goals, objectives, criteria, and performance requirements will already have been addressed and ordinarily need not be part of the analysis.

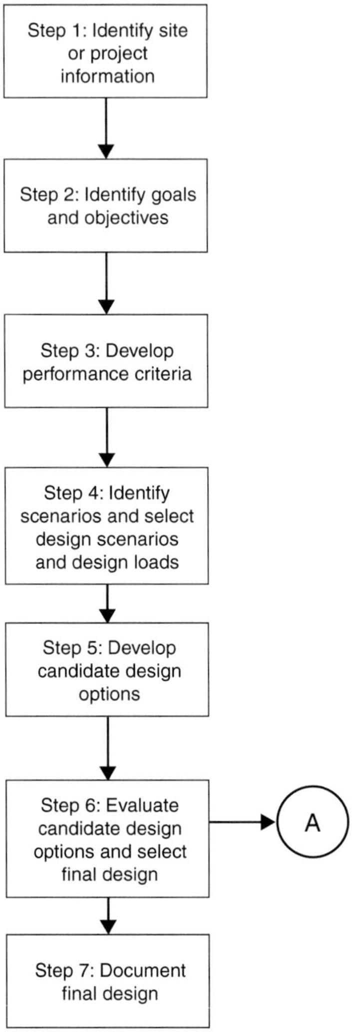

FIGURE 2.2 Steps in a performance-based analysis and design process.

Verification Methods

Verification methods can be defined to include test standards; test methods; and analytical methods, including computational models. In short, a verification method can be any document, system, test, method, or tool that is needed to measure a design and its components against goals, functional statements, operational requirements, and performance criteria.

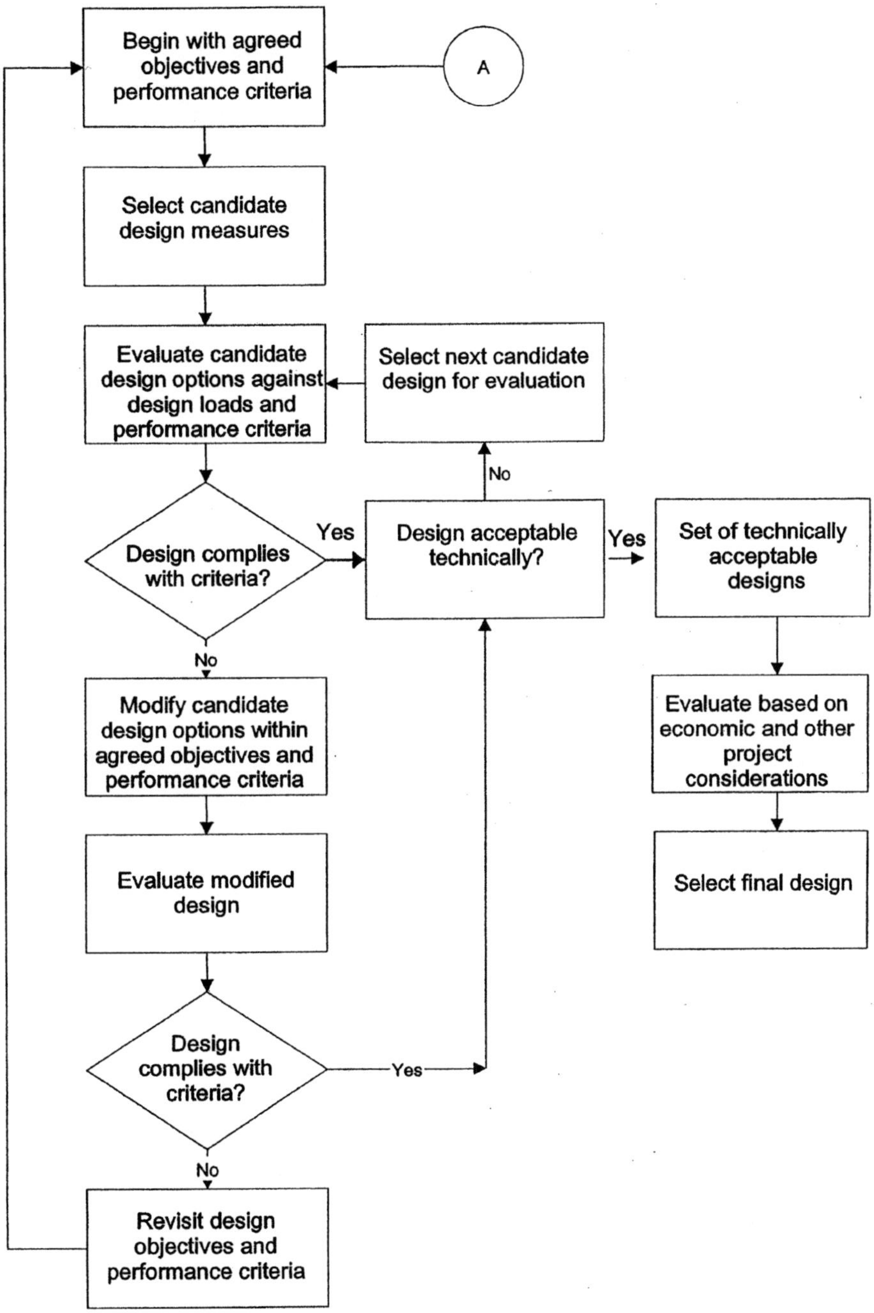

FIGURE 2.3 Iteration in the performance-based design process.

In a performance-based system, verification methods are often computer-based analytical tools that require a combination of data and judgment, because it may not be cost-effective or even possible to physically test every possible mitigation solution. In these cases, the tools and the data needed for the analysis must be suitably validated, and uncertainty, variability, and unknowns appropriately accounted for. Personnel must be qualified to understand the totality of a risk-informed performance-based approach when they use such tools. Otherwise, there is a risk that one area may receive a disproportionate amount of analysis to the detriment of overall safety or of the facility as a whole.

SUMMARY

This chapter provides an overview of a performance-based approach to regulation and design of buildings, with a focus on security and blast mitigation. It illustrates how risk and protection can be used as performance metrics, and how performance-based and prescriptive criteria can be used jointly in design guidance. The committee believes that it can serve as a guide for the Interagency Security Committee to use in developing a performance-based process for the security-related design of federal facilities.

REFERENCES

Custer, R.L.P., and B.J. Meacham. 1997. Introduction to Performance-based Fire Safety. Boston, Mass.: Society of Fire Protection Engineers.

DOJ (Department of Justice). 1995. Vulnerability Assessment of Federal Facilities. Washington, D.C.: U.S. Department of Justice.

Hamburger, R.O., A.B. Court, and J.R. Soulages. 1995. Vision 2000: A framework for performance-based engineering of buildings. Proceedings of the 64th Annual Convention, Structural Engineers Association of California, pp. 127-146.

ICC (Interstate Commerce Commission). 2001. International Performance Code for Buildings and Facilities. Falls Church, Va.: ICC.

ISC (Interagency Security Committee). 2001. ISC Security Design Criteria for New Federal Office Buildings and Major Modernization Projects. Washington, D.C.: GSA.

Meacham, B.J. 1998a. Concepts of a Performance-Based Building Regulatory System for the United States. NIST-GCR-98-762. Gaithersburg, Md.: National Institute of Standards and Technology.

Meacham, B.J. 1998b. Assessment of the Technological Requirements for the Realization of Performance-Based Fire Safety Design in the United States. NIST-GCR-98-763. Gaithersburg, Md.: National Institute of Standards and Technology.

Snell, J. 2001. Towards a global agenda for fire research. Proceedings of the United Engineering Foundation Conference: The Technical Basis for Performance-Based Fire Regulations. New York, N.Y.: United Engineering Foundation.

3
Blast Effects on Buildings and People: A Primer for Users of the ISC Security Design Criteria

In addition to natural and technological hazards, in recent years owners, tenants, and design professionals have had to consider design criteria to address an additional hazard to public buildings—the deliberately placed bomb. Whether a new federal facility is being built or an existing one renovated, physical protection from bombs has become integral to the planning process. The challenge for the building project team is to design and construct federal facilities that provide protection from terrorist explosive threats while at the same time offering desirable workspaces in attractive buildings that are well integrated with the surrounding neighborhood. When these structures are situated on urban sites, it is often difficult to restrict access to preserve effective standoff[1] distances. Architectural aesthetics may also conflict with blast mitigation. Given the limited resources available for physical protection, these competing objectives are often addressed through design solutions that seek to protect occupants while accepting significant damage to the assets and structure. There is a need for a better way of allocating scarce resources to the highest-priority threats.

Comprehensive protection against the full range of possible threats is prohibitively costly—and probably not even possible. However, the committee believes that a level of protection that reduces the risk of mass casualties resulting from terrorist attacks can be provided for all occupants

[1]*Standoff* (sometimes referred to as *setback*) is generally understood to be the distance between the detonation point of a bomb and the target building.

at a reasonable cost. Full implementation of the ISC Security Design Criteria will provide some protection against most threats while significantly reducing the risk of injuries.

PURPOSE AND USE

The many possible users of the ISC Security Design Criteria will have differing degrees of familiarity with the topics addressed; in general, the audience may require an introduction to the subjects of both blast effects on people and buildings and the technical basis for blast-resistant design. Among those who should benefit from a general description of the basic concepts of blast-resistant design are architects and engineering professionals, facility owners, agency and private project managers, and tenants and other users of the facilities. Therefore, this basic introduction to protective design is provided to supplement the technical discussions of other parts of the ISC Security Design Criteria.

Understanding both features for physical protection of buildings and operational security features can inform decisions on physical protection. This explanation may help dispel some misconceptions about protective design and elicit more cooperative participation by all stakeholders in implementing effective design strategies from the outset. Stakeholders can readily see the benefits of such basic ways to protect buildings from bomb damage as establishing a secure perimeter, preventing progressive collapse, isolating internal threats from occupied spaces, and mitigating the glass and debris hazard.

Improving the performance of a building in response to an explosive event requires the services of a professional experienced in both conventional and protective design. First, qualified security professionals, in consultation with the building owner, manager, and occupants, should assess how vulnerable a facility is to various types of threats and assign levels of risk. A threat assessment and risk analysis (TARA) can help guide the owner, principal tenants, and security staff in balancing the operational, technical, and physical aspects of security services to maximize protection within the budget available.

BLAST EFFECTS—BASIC INFORMATION

The Nature of Explosions

Explosive materials are designed to release a large amount of energy in a very short time. Part of the energy is released as heat and part as shock waves that travel through the air and the ground. The shock wave (air blast) radiates at supersonic speed in all directions from the explosive source,

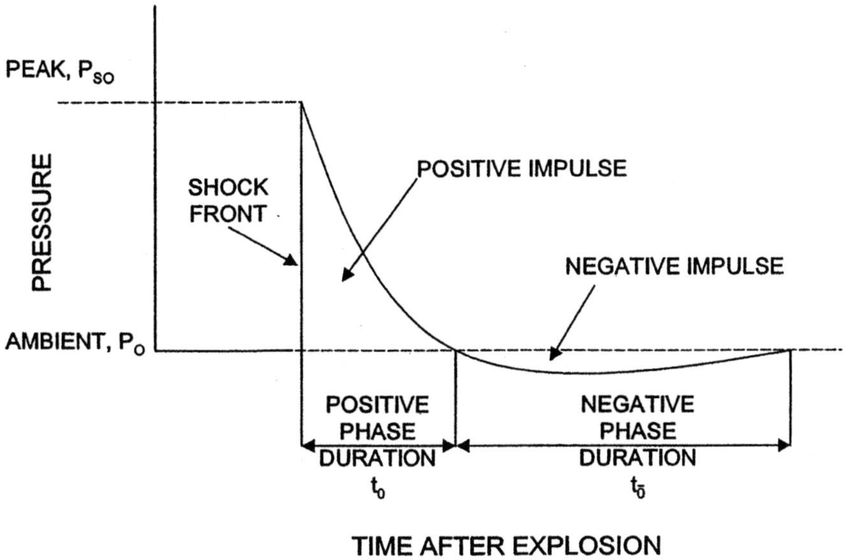

FIGURE 3.1 Pressure-time relationships after an explosion.

diminishing in intensity as the distance from the source increases. These waves reflect off the ground, adjacent structures, and other surfaces, reinforcing the intensity of the blast's effects. These reflections are most pronounced in dense urban environments where neighboring structures can create a canyon effect or where pockets of blast energy can become trapped in re-entrant (concave) corners.

Blast wave reflections also occur within structures, greatly increasing the potential for injury to the occupants and damage to building systems and contents. When a shock wave encounters a structure, it subjects all surfaces to the force of the blast. If the shock wave enters through windows or door openings, the interior elements—both structural (e.g., floor slabs and columns) and nonstructural (partitions, lighting fixtures, etc.)—the contents, and the occupants will also be subjected to the effects of the airblast. Moreover, there are dynamic pressures that occur somewhat later caused by the mass movement of the air itself. This is essentially a high-velocity wind that can propel blast-generated debris with great force.

In very large explosions, the shock wave propagated through the earth may induce ground motion similar to that resulting from a short-duration earthquake. As shown in Figure 3.1, immediately after an explosion the pressure increases to a peak value almost instantaneously. As the pressure then decays from its peak value and interacts with the structure, it can cause enormous damage. Impulse (a blast parameter of momentum) is the time during which the pressure wave decays back to its pre-explosion ambient value; it is represented by the area under the pressure-time curve. For high explosives, the shock wave traverses the structure very quickly and the positive-phase duration, t_0, is measured in milliseconds. Nuclear weapons have a much longer positive-phase duration, which partially explains their capacity to cause far greater structural damage.

The structural and other damage caused by an explosion is the building's response to the enormous amount of energy produced. This energy can be either resisted through the use of massive elements that are strong or ductile enough to survive without failure or accepted in the form of partial damage to windows, facade, and structural members. The concept of "graceful failure" assumes that various elements will resist long enough to absorb a large amount of energy and then fail in a way that minimizes the risk of serious injury or death to the occupants. Modern protective design, which attempts to balance security and aesthetics, generally incorporates aspects of both initial resistance and graceful failure.

Blast Effects on Buildings

Buildings experience the effects of explosions in several stages:

- The initial blast wave typically shatters windows and causes other damage to the building facade. It also exerts pressure on the roof and walls that are not directly facing the blast, sometimes damaging them as well.
- In the second stage, the blast wave enters the building and exerts pressure on the structure. When directed upward, this pressure may be extremely damaging to slabs and columns because it acts counter to the design used to resist gravity loads. Air-blast pressures within a building can actually increase as the pressure waves reflect from surfaces and can cause injuries to the occupants directly by means of physical translation, ear, lung, and other organ damage, or debris from building elements and contents.
- Finally, the building frame is loaded globally and responds as it would to a short-duration, high-intensity earthquake.

The blast pressures experienced by a structure are in a most general sense related to the amount of explosive used and the distance of the building from the explosion. The peak incident pressure, charge weight, and distance are mathematically related through an expression that varies as a function of the weight of the explosive and the cube of the distance. This relationship is critical to understanding the effects of explosions on structures. In particular, the pressure experienced by a building increases

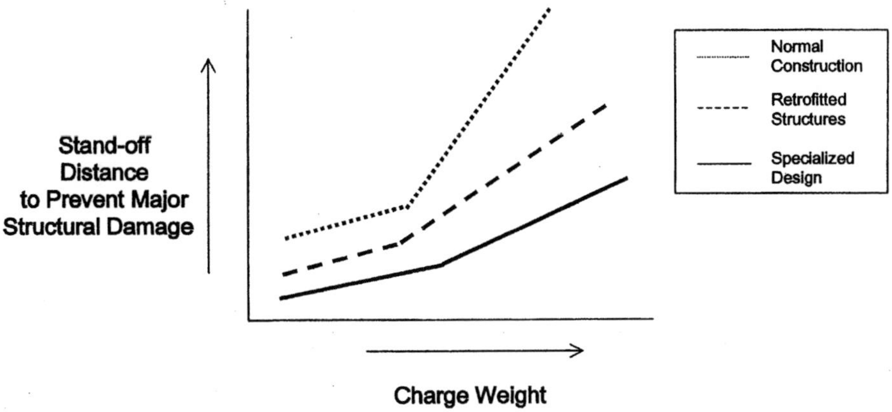

FIGURE 3.2 Effect of standoff distance on building protection requirements (based on Jenssen, 1998).

with bomb size, but decreases very quickly with increasing distance between the building and the bomb. These factors form the two keystones of defensive design and, as is shown in subsequent discussion, limiting the size of a bomb through vehicle control and inspection and enforcing standoff distances from possible targets are two of the most important tools available to those charged with protecting people and buildings from bomb damage. Figure 3.2 illustrates how standoff distance can prevent major structural damage for typical commercial construction, structures retrofitted to increase blast resistance, and specially designed blast-resistant structures.

Usually, the standoff distance available and the assumed size of the explosive device will determine the blast-resistant features that must be provided. Large explosive devices detonated at relatively great standoff distances will produce a large but uniform pressure over the surface of the building; at lesser standoff distances, even a small explosive device can produce locally intense effects, such as shattering load-bearing columns. While the former scenario is likely to govern design of the facade to limit the formation of hazardous debris, the assumption of a smaller, close-in device is likely to control design of the first-floor load-bearing elements to prevent localized failure leading to progressive structural collapse.[2] If a large explosive device is detonated close to the structure, global damage

[2]Progressive collapse occurs when a localized failure overloads adjoining members, causing them to fail, which in turn causes damage disproportionate to the originating localized failure.

and the size of the resulting ground crater may be increased to the point that the structure, foundation, or both will be overwhelmed and a catastrophic collapse may ensue.

One of the important first steps in the design of any structure is defining the loads. In security design, this must include defining the blast loading—a function of the expected charge weight and proximity. Although a vehicle-bomb employing conventional explosives still appears to be the most serious bombing threat, experience over the past decade has shown that hundreds of smaller bombing incidents occur annually (ATF, 1996). Although these smaller bombs lack the power to cause catastrophic structural damage, they can injure and kill people and cause localized damage, particularly to windows and nonstructural elements. To provide a proper basis for risk assessment and cost-benefit evaluation, the information used to define blast loads or expected charge weights must be as unambiguous as possible. This can best be accomplished by carefully analyzing charge weights from prior incidents, defining their probability distribution properly, and selecting design blast loads at specified probabilities.

Blast Effects on People

People exposed to explosions can be killed or injured by the intense heat and pressure generated at the site of the detonation, where temperatures can range up to 4,000°C and pressures to several hundred times atmospheric. Such extremely high pressures can damage major organs, blood vessels, eyes, and ears. Shock-tube and explosive tests have indicated that human blast tolerance varies with both the magnitude of the shock pressure and the shock duration; the pressure tolerance for short blast loads is significantly higher than for long blast loads (U.S. Department of the Army, 1990).

Once the blast wave enters occupied spaces or in an indoor explosion, the pressure wave is reflected off walls, floors, and ceiling, forming in effect a series of pressure pulses. The response of the ears and lungs to repetitive pulses is similar to the response to a single long pulse. Because the damaging effects of blast pressure indoors often exceed those of an unconfined explosion of similar size, it is crucial to minimize the opportunity for blast pressures from outdoor explosions to enter occupied spaces and to protect such spaces from even small explosive devices (Cooper, 1996). How the ear and lung respond depends on the impulse and the orientation of a person's body to the blast wave—the shorter the pulse duration, the higher the pressure that can be tolerated. The onset of lung hemorrhage begins in the range of 30 to 40 psi, with severe damage occurring above 80 psi and death in the range of 100 to 120 psi. The onset of eardrum rupture is about

5 psi; 50 percent of eardrums exposed to a rapidly rising pressure of 15 psi will rupture (U.S. Department of the Army, 1990).

Medical reports of past bombings and recent suicide bombing attacks cite shock and organ trauma as leading causes of death (Brismar and Bergenwald, 1982; Rivkind, 2002; Weightman and Gladish, 2001). For people within structures subjected to blast effects, penetration by glass fragments and impact by other blast-induced debris have been consistent causes of death and serious injury (Mallonee et al., 1996). This underscores the need for window treatments that minimize the production of fragments and for catcher systems to retain debris. People are also subject to blunt trauma from furniture, accessories, and nonstructural building components like overhead lighting and ductwork that become detached from their moorings. Smoke and inhalation of dust also cause blast-induced injury. That is why it is important to make every effort to keep blast energy outside the building and to secure furniture and nonstructural components. Earthquake hazard mitigation offers practical guidance here (FEMA, 1994). Numerous injuries are often sustained during evacuation, particularly if there is a large amount of shattered glass littering evacuation routes or if there is considerable smoke and dust.

Analysis of the 1995 Oklahoma City bombing (Mallonee et al., 1996) demonstrated that building collapse was the primary cause of death. In the 1996 bombing of the Khobar Towers residential complex in Saudi Arabia, the structure did not collapse; although the number of fatalities was far less than in Oklahoma City, their causes were similar to those in Oklahoma City that did not result directly from the building collapse (Oklahoma State Department of Health, 2000). Studies of the attacks on the U.S. embassy in Nairobi also found that the harm can be substantial even if the structure does not collapse. The board that reviewed the embassy bombings in Nairobi and Dar es Salaam in 1998 found:

> The damage to the embassy was massive, especially internally. Although there was little structural damage to the five-story reinforced concrete building, the explosion reduced much of the interior to rubble—destroying windows, window frames, internal office partitions, and other fixtures on the rear side of the building. The secondary fragmentation from flying glass, internal concrete block walls, furniture, and fixtures caused most of the embassy casualties. (U.S. Department of State, 1999)

The Terrorist Threat

Although there is no theoretical limit to the size of an explosive device or the locations where it might be placed, there are practical implications of size and weight for explosive devices that could constitute terrorist weapons. The density of common high-energy explosives, such as ammonium

TABLE 3.1 Explosive Capacity of Typical Bomb Delivery Methods

Delivery Method	Explosive Capacity (pounds/kilograms)
Mail bomb	5/2.3
Suitcase bomb	50/23
Automobile	500-1,000/225-450
Van	4,000/1,800
Truck	10,000-30,000/4,500-13,500
Semitrailer	40,000/18,000

SOURCE: TSWG (1999).

nitrate and fuel oil (ANFO), TNT, and C4, is on the order of 100 pounds per cubic foot. The explosive capacities of various delivery devices ranging from a small suitcase bomb to a large truck can therefore be calculated: 50 pounds may be packaged in a large briefcase that fits unobtrusively on a luggage carrier. Smaller amounts may be packaged in backpacks and shopping bags, all of which arouse little suspicion from experienced security guards. Explosive devices weighing anywhere from several ounces to several pounds may be delivered through the U.S. Postal Service, other delivery services, or private couriers. Each of the several different sizes of the blast events typically considered in a terrorist threat analysis corresponds to the type of container or vehicle in which the explosive is delivered (see Table 3.1).

PROTECTIVE DESIGN STRATEGIES

Overview

Physical protection for buildings involves four basic actions:

- Establish a secure perimeter.
- Prevent progressive collapse.
- Separate internal threats from occupied spaces.
- Mitigate debris resulting from damage to the facade and building interior.

Other considerations, such as anchoring nonstructural components inside the building, are also design objectives that require special attention, as is protection of emergency services. The size of the explosive threat will determine the effectiveness of each of these protective features and the resources necessary to protect building occupants. Defining the threat to be designed against is fundamental to the protective design process and therefore requires careful consideration.

Perimeter Control

Though definition of the design threat depends on history, available intelligence, and assumed risk, from a practical standpoint it is limited by the means of weapon delivery. Because conventional explosives weigh approximately 100 pounds per cubic foot, a hand-carried device, if efficiently packaged, could occupy as little as half a cubic foot of space in a large briefcase or small suitcase. Such a device may be introduced deep into the structure, where it can do considerable damage to structural members, such as load-bearing columns, or cause considerable injury to occupants. As a result, screening stations at public entrances, mailrooms, and loading docks are critical for preventing these threats from entering occupied space.

Vehicles can carry significantly larger explosive charge weights. As a result, perimeters must be secured and parking or loading docks underneath or within occupied buildings must be protected by comprehensive screening procedures, performed continuously. The potential for threats to bypass screening procedures will always exist. Therefore, the choice of design-level explosive threat will be guided by standoff distance, site conditions, blast-resistant features of the building, and ultimately the amount of risk the building owner or principal tenant is prepared to accept.

Although it is theoretically possible to predict the effects of a certain charge weight of a known explosive at a specified standoff distance, the actual charge weight of explosive used by a terrorist, the efficiency of the chemical reaction, and the location cannot be reliably predicted. Thus the approach embodied in the ISC Security Design Criteria is to use predefined levels of protection based on an estimation of risk to the facility, taking into account its symbolic importance, mission criticality, and the consequences of loss.

Although this approach is a start for risk analysis, once a facility is perceived to be at risk of attack, initial assumptions about charge weight can be based on the capacity of possible delivery vehicles. In light of the large capacity of most trucks and vans and the practical limitations of structurally hardening general-purpose buildings to resist the effects of very large bombs, the most effective way to protect a structure is to keep the bomb as far away as possible—maximize the standoff distance.

To guarantee the integrity of the perimeter defense, antiram bollards, large planters, or other devices can be placed around the site at the desired standoff distance. Site conditions will govern the maximum vehicle speeds attainable. Both the barrier and its foundation must be designed to resist the maximum energy that can be developed by the delivery vehicle. If conditions limit the capacity of the barrier or its foundation, other means must be used to reduce vehicle speed or size.

Parking abutting the building should be secured or eliminated; uncontrolled street parking should not be permitted near the building. Though converting one lane of traffic into an extended sidewalk or plaza can increase standoff distance, the practical benefit of increasing the standoff depends on the charge weight. If the charge weight is small, even small increases in standoff distance will significantly reduce blast forces, but if the threat is large, the blast forces may overwhelm the structure despite the addition of 9 or 10 feet of standoff distance, and this measure may not significantly improve the survivability of the occupants or the structure.

Glazing Systems

The building exterior is the first real defense against the effects of a bomb; how the facade responds will significantly affect the behavior of the structure and the safety of its occupants. Hardening the facade is typically the single most costly and controversial component of blast protection because it is likely to change the appearance of the structure dramatically. The number and sizes of windows usually will be reduced and their attachments to the structure will become more rugged. Considering the large surface areas enclosing some buildings, the cost for increasing glazing system protection even modestly will be high. As a result, the optimal design solution is often one that seeks to improve how the facade behaves after some specified degree of damage.

Window assemblies can be designed to respond to many anticipated blast loads. For new construction, it may be best to specify laminated glass; for existing glazing, a fragment-retention film might be applied. Although these approaches do little to improve the strength of the glass (failure *will* occur), they can hold the shards of glass together to better protect occupants from hazardous debris.

Window solutions must be approached cautiously, however. There is a danger that strengthening glass could cause the entire window light to be dislodged in a single piece, injuring or killing anyone struck by it. Catchbar and other systems can provide additional safety if they are aesthetically and functionally viable, but ultimately, the risk of injury posed by many small shards must be weighed against that posed by a single large sheet.

The effectiveness of fragment-retention films depends on how they are applied, the thickness of the film, and how it is anchored to the frame. Common film systems range from simple edge-to-edge (daylight) applications to wet-glazed adhesion to a mechanical attachment to the window frame. Mechanical attachments are most effective when they are anchored to the underlying structure. Regardless of the method used, however, fragment-retention films raise architectural and life-cycle cost issues. Laminated glass is another option. It exhibits excellent post-damage behavior, is available for most applications, and provides a high degree of safety to occupants.

The design of the window frames is as important as the type of glazing material used. If the window system is to fail safely, the glass must be held in the frame long enough for the stresses to cause it to fracture. Otherwise, the intact glass panel could separate from the frame at high speed and cause serious injury. Therefore, the frame system should be designed to allow the full mechanical capacity of the chosen glazing material to develop. Because the nominal strength of glazing is specified in terms of a prescribed number of failures per thousand, to ensure that the frames can resist the full mechanical capacity of the glass, a higher strength for the glazing material must be considered. Factors of two or three times the nominal capacity of the glass to resist breakage may be the basis for frame design. The bite (the distance the glass extends into the frame) must be adequate to assure that the failed glass is retained within the frame—possibly using structural silicone sealant. Finally, depending on the facade, the mullions may be designed to span from floor to floor or to tie into wall panels. They must also be able to withstand forces that will cause the window to fail.

Curtain Walls

A curtain wall is a nonbearing exterior enclosure that can hold glass, metal, stone, precast concrete, or other panels and is supported by a building's structural frame. Lightweight and composed of relatively slender members, curtain walls require careful design and considerable testing to ensure that the assembly has adequate wind and water resistance and meets the light and temperature differentials required of the exterior envelope. The ability of a curtain wall system to withstand the effects of explosive loading depends on how the various elements of the system perform. Although the glazing may be the most brittle component of a curtain wall system, the risk to the occupants depends on how the mechanical capacities of its elements interact. In addition to hardening[3] the members that compose a curtain wall system, attachments to the floor slabs, or spandrel beams, require special attention. These connections must be adjustable to compensate for fabrication tolerances, accommodate differential interstory drifts and thermal deformations, and yet be capable of transferring gravity, wind, and blast loads.

An alternative approach is to allow the window systems to absorb a

[3] Blast-hardening of a structure refers to all measures taken, either in the design phase or in retrofit actions, to reduce or eliminate the effects of an explosion. This process is sometimes referred to simply as building "hardening" (NRC, 1995).

considerable amount of blast energy through deformation while preventing debris from entering the occupied space. Explosive tests have determined that the inherent flexibility of curtain wall systems allowed their glazing to survive higher blast pressures than rigidly supported windows. The cable-protected window system takes this concept one step further; as the glass is damaged, it bears against a cable or muntin catch system, which in turn deforms the window frames. This system makes full use of the flexibility and capacity of all the window materials and dissipates large amounts of blast energy without hurling debris into the protected space. Extensive explosive testing has demonstrated the effectiveness of this approach, as have sophisticated computer simulations.

Structural Issues

In addition to facing the hazard of glass and other facade debris being propelled into a building, its occupants may also be vulnerable to injury from much heavier debris resulting from structural damage. When an initiating localized failure causes adjoining members to be overloaded and fail, this progressive collapse causes damage that is disproportionate to the originating localized failure. A protective design will not use structural systems that either facilitate or are vulnerable to progressive collapse. In particular, new facilities may be designed to accept the loss of an exterior column for one or even two floors above grade without precipitating collapse.

These design requirements are intended to be threat-independent, providing redundant load paths should any damage occur due to abnormal loading. Threat-independence is intended to protect against an explosion of indeterminate size that might damage a single column. Although the alternate path approach is not associated with any specific threat that might cause damage, it is limited to abnormal loading conditions that would fail only one load-bearing member. Upgrading existing structures to prevent localized damage from causing a progressive collapse may not be easy using the alternate path method because loss of support at a column line would increase the spans of all beams directly above the zone of damage and require different patterns of reinforcement and different types of connection details than those typically used in conventional structural design.

Alternatively, columns may be sized, reinforced, or protected to prevent critical damage from a nearby bomb. Vulnerable concrete columns may be jacketed with steel plate or wrapped with composite materials. Steel columns may be encased in concrete to protect their cross sections and add mass. These approaches to preventing progressive collapse are generally more feasible in retrofits than attempting to supplement the capacity of connecting beams and girders; however, their effectiveness is predicated on operational and technical security procedures that limit the magnitude and proximity of an explosive threat. These include effective perimeter protection, adequate screening of vehicles entering a parking facility or loading dock underneath or within the building, limiting parking adjacent to the building, and inspecting parcels carried into the building.

Transfer girders and the columns supporting them are particularly vulnerable to blast loading. Because transfer girders typically concentrate the load-bearing system into fewer structural elements, they do not provide the redundancy desired for blast resistance. Typically, the transfer girder spans a large opening, such as a loading dock, or makes it possible to shift the location of column lines at a particular floor. Damage to the girder may leave totally unsupported several lines of columns that terminate at the girder from above. The loss of a support column from below will also create a much larger span to carry critical load-bearing structure.

Transfer girders are critical sections; their loss may result in progressive collapse. If a transfer girder must be used, and if this girder may be vulnerable to an explosive loading, it is desirable that it be continuous over several supports. Furthermore, there should be substantial structure framing into the transfer girder to create a two-way redundancy and an alternate load path if there is a failure.

Other Building Spaces and Systems

The walls surrounding loading docks, mailrooms, and lobbies into which explosives may be introduced before inspection must be hardened to confine an explosive shock wave and must permit the resulting gas pressures to vent into the atmosphere. The isolation of occupied spaces from these vulnerable locations requires adequate reinforcement as well as connections that can resist the collected blast pressures. These structural features must be well integrated with the rest of the structural frame to ensure that any failure does not destabilize other portions of the gravity load-bearing system.

Because a significant area of the exterior walls enclosing mechanical spaces is louvered for ventilation, hardened plenums are required to interdict a direct line between the louvered opening and the machinery beyond. The response of electrical and mechanical equipment to blast pressures has been studied using lethality and survivability algorithms developed for the Department of Defense. In the design process, all components composing the critical electrical and mechanical systems must be analyzed in detail to determine whether infill blast pressures or in-structure shock motions would disrupt service.

Nonstructural building components, such as piping, ducts, lighting units, and conduits, must be sufficiently anchored to prevent failure of services and so that they do not become falling debris. To mitigate the effects of shock due primarily to the entry of blast pressures through damaged windows, these nonstructural systems should be located below raised floors where possible, or tied to the ceiling slabs using restraints appropriate for Seismic Zone IV (FEMA, 1994).

APPROACHES TO BLAST-RESISTANT STRUCTURAL DESIGN

Dynamic Versus Equivalent Static Analysis Methods

The pressure waves that result from detonation of an explosive device have a very high intensity peak value that diminishes to zero in milliseconds. Both the intensity of the blast pressures and their duration greatly influence their effect on structures. The relation between the brevity of the loading and the natural period of response of individual structural elements, as well as the ability of these structural elements to deform inelastically without collapse, will determine resistance. Massive structural components provide inertial resistance, which tends to reduce the magnitude of resistance required. While the strength of these members is critical to their response, their ability to deform inelastically in a ductile fashion will limit the forces that must be resisted. These effective loads may be fractions of the peak blast pressure intensity; blast analysis for a particular structural component will determine the required resistance. The design and detailing of structural elements make it possible for the structure to deform in a ductile manner to prevent a catastrophic brittle failure and allow for timely evacuation of the facility. Although significantly reduced from the peak intensity of blast pressures, these loads may still be many times greater than the design loads required for gravity, winds, or even seismic disturbances.

Seismic Versus Blast-resistant Design

Although design for seismic resistance and design for blast resistance share some common principles, the two types of design must not be mistakenly viewed as redundant. Unlike a seismic disturbance, in which the induced forces are proportional to the distributed mass, blast loading does not tax an entire frame uniformly. The localized loads will deform exterior bays of the structure much more than interior bays; as the blast loading progresses, the shear forces in each story of the building may not necessarily be distributed through the diaphragms in proportion to the framing stiffness. Furthermore, the characteristic patterns of loading and deformation in a blast event depend to a great extent on the standoff distances, which may be significantly different from those resulting from a seismic excitation of the structure. Finally, if the initiating damage is widespread, the redistribution of forces to the already weakened structure may amplify the extent of the collapsed region; the resulting structural failure will be a result of global damage, not progressive collapse.

It is generally understood that increasing the ductile behavior of details in response to strong ground motions will increase the ductile behavior in response to blast loading. Yet though seismic design codes contain very useful detailing information that may be directly applied in blast resistance design, it is generally acknowledged that the zones of plastic hinge formation and the extent of ductility demands for seismic response are not necessarily useful. Attempts to link seismic and blast design requirements by simply comparing the lateral or shear forces on a structure produced by these events (an equivalent seismic base shear) perpetuate the erroneous impression that seismic design is an umbrella for blast resistance.

Assuming the initiating failure is localized to a single column, it is desirable and entirely possible to design a new structure to limit the extent of collapse to the bays on either side of the failed column for a height of one floor. The alternate path method, which assumes an idealized zone of initiating failure, allows the engineer to quantify the amount of continuously tied reinforcement that should be detailed into concrete members or steel sections. However, the zone of initiating failure assumption is an academic idealization. If the zone of damage is more extensive, the calculation of force redistribution will be in error; the definition of disproportionate collapse must be related to the entire region damaged by the explosion. Except for the near-contact satchel-charge scenario, the actual extent of damage may be significantly greater than the removal of a single column, and so the definition of disproportionate collapse must be revised to reflect this reality.

Furthermore, progressive collapse must not be confused with global collapse of structural systems. Progressive collapse cannot be prevented if the blast is so intense as to damage significant portions of the structure simultaneously. This would be termed a general collapse, and the alternate path approach will provide only limited resistance to global damage mechanisms. Localized hardening of vulnerable structural elements and improving robustness through ductile detailing of structural systems will improve resistance to more extensive blast damage.

REFERENCES

ATF (Bureau of Alcohol, Tobacco, and Firearms). 1996. Explosives Incidents Report. Washington, D.C.: U.S. Department of the Treasury.

Brismar, B., and L. Bergenwald. 1982. The terrorist bomb explosion in Bologna, Italy, 1980: An analysis of the effects and injuries sustained. Journal of Trauma 3:216-220.

Cooper, P.W. 1996. Explosives Engineering. New York, N.Y.: Wiley-VCH.

FEMA (Federal Emergency Management Agency). 1994. Reducing the Risks of Nonstructural Earthquake Damage: A Practical Guide. Washington, D.C.: Federal Emergency Management Agency.

Jenssen, A. 1998. Infrastructure protection in the United States: A Norwegian perspective. Presentation at the Workshop on the Use of Underground Facilities to Protect Critical Infrastructure. National Academy of Sciences. Washington, D.C., April 6-7, 1998.

Mallonee, S., S. Shariat, G. Stennies, R. Waxweiler, D. Hogan, and F. Jordan. 1996. Physical injuries and fatalities resulting from the Oklahoma City bombing. Journal of the American Medical Association 5:382-387.

NRC (National Research Council). 1995. Protecting Buildings from Bomb Damage: Transfer of Blast-effects Mitigation Technologies from Military to Civilian Applications. Washington, D.C.: National Academy Press.

Oklahoma State Department of Health, Injury Prevention Service, and the U.S. Air Force. 2000. An Epidemiologic Investigation of Physical Injuries Resulting from the Khobar Towers Bombing, Final Report. Prepared under contract for the U.S. Department of Defense, Technical Support Working Group, Arlington, Va., September 25, 2000.

Rivkind, A. 2002. A doctor's story: Awaiting the wounded. Chicago Tribune, July 14.

TSWG (Technical Support Working Group). 1999. Terrorist Bomb Threat Standoff. Washington, D.C.: Technical Support Working Group.

U.S. Department of the Army. 1990. Structures to Resist the Effects of Accidental Explosions. Army TM 5-1300, Navy NAVFAC P-397, AFR 88-22. Washington, D.C.: Departments of the Army, Navy, and Air Force.

U.S. Department of State. 1999. Report of the Accountability Review Boards on the Embassy Bombings in Nairobi and Dar es Salaam on August 7, 1998, January 1999. Available online at <http://www.state.gov/www/regions/africa/accountability report.html>.

Weightman, J., and S. Gladish. 2001. Explosions and blast injuries. Annals of Emergency Medicine 6:664-678.

4

Assessment of the ISC Security Design Criteria

THE ISC SECURITY DESIGN CRITERIA

The *ISC Security Design Criteria* document (ISC, 2001) is a mix of performance objectives, prescriptive requirements, and references to industry design standards. The design criteria are generally acceptable and full implementation of the ISC Security Design Criteria will provide some protection for building occupants against most blast-related threats and will significantly reduce injuries. However, the organization of the document makes it difficult to identify clearly the connections between specific criteria and the performance objectives they are meant to achieve.

The document is also focused on the terrorist vehicle bomb as the primary means of attack; there is little guidance on defending federal buildings and their occupants from chemical, biological, or radiological weapons.

Without overarching goals and objectives and without clearly illustrated connections between desired performance and proposed solutions, the document encourages consideration of specific aspects of building design in isolation (e.g., structure independent of fire protection). This could result in building performance not being considered comprehensively—for instance, failure to design for survivability of fire protection systems after a bomb attack could result in an otherwise avoidable post-attack fire.

There is also a lack of consistency in the document's language, grammatical structure, and flow that makes it difficult to comprehend as a whole. This could confuse those responsible for the physical security component of building design, as well as those who must review designs and make decisions about design options. These and related issues can be addressed by continuing refinement of the *ISC Security Design Criteria,* clarifying and reformatting the document to provide both more complete and comprehensible design criteria and the necessary policy guidance.

The continued use of both prescriptive and performance criteria is appropriate for several reasons, including the fact that in much of the building design process using prescriptive criteria need not limit creative design. Performance analysis and design are only needed for certain portions of the process, for example, the design of glazing to satisfy a unique threat or location. Structured appropriately, prescriptive criteria can be a means of meeting performance objectives. However, the ISC Security Design Criteria do not provide guidance on the amount and completeness of information to be provided in documenting a performance-based security design. Documentation is a way of verifying that scenarios of good design practice have been analyzed and applied. It should demonstrate compliance with performance objectives, review and acceptance by trained government officials or their designees, construction inspection and testing, and issuance of the documents necessary to demonstrate that occupancy and related requirements have been met.

Responsibility for Implementing and Overseeing the ISC Criteria

Executive Order 12977 (Clinton, 1995) established the Interagency Security Committee (ISC) as "a permanent body to address continuing government-wide security." The ISC has a broad membership of 17 federal agencies. Although the ISC has produced the security standards mandated in Section 5, items (1) and (2) of the Executive Order, it does not appear to be overseeing "the implementation of appropriate security measures"; nor has the General Services Administration (GSA) Administrator put in place the means to monitor "Federal agency compliance with policies and recommendations of the Committee [ISC]," as mandated by Section 6, paragraph (c).

By asking the committee to evaluate the ISC Security Design Criteria to determine if they were prescriptive or performance-based, it is understood that GSA desires to give as much freedom as possible to designers of its facilities by allowing them to use performance-based rather than prescriptive—and presumably more restrictive—criteria wherever possible. The committee believes that there are two organizational constraints to achieving both the intent of Executive Order 12977 and the objective of allowing designers maximum freedom to apply imaginative solutions. First, there must be a professional staff with the qualifications and authority to evaluate and approve proposed solutions. Second, that staff must be housed within an agency with the commitment and stability to see the program through.

Outsourcing Management Functions for the Acquisition of Federal Facilities (NRC, 2000) describes a "smart owner" in the commercial design, engineering, and construction industry as a business entity that has the skill base—usually a staff with professional qualifications and authority—necessary to plan, guide, and evaluate the facility acquisition process. A smart owner asks how a specific facility will contribute to the success of the organization's mission. This has numerous implications for a performance-based regulatory approach, including increased responsibility for building owners, design professionals, and approval authorities, as well as training and education issues, qualification and competency requirements, facilities management issues, and financial implications. Whether security-related design remains a responsibility of the ISC as provided in Executive Order 12977 or devolves to a more central or security-focused entity, such as the Department of Homeland Security, the government will need to meet the definition of a smart owner if it is to implement a performance-based security design program properly.

GSA Facility Standards

In its review of the ISC Security Design Criteria, the committee also considered the *Facilities Standards for the Public Buildings Service* (GSA, 2000). Available on the Internet and in hard copy, the *Facilities Standards* is GSA's attempt to apply the ISC Security Design Criteria in its role as the federal government's largest provider of space and facilities to federal agencies for nonmilitary users. The *Facilities Standards* is issued to architects and engineers who design buildings for GSA's Public Buildings Service. However, the committee finds it troublesome that the *Facilities Standards* document does not include the text from Part 1 of the *ISC Security Design Criteria* that would connect the criteria, in a policy sense, with the Presidential Executive Order that established the Interagency Security Committee. Absent this linkage or another official promulgation, there is no evidence that GSA requires its regional offices to apply these standards.

The committee assumes that Executive Order 12977, which created the ISC and mandated enhancement of "the quality and effectiveness of security in and protection of buildings and facilities in the United States occupied by Federal employees," continues to be the driver of the measures contained in the *ISC Security Design Criteria* and Chapter 8 of the *Facilities Standards*. Without the guidance provided by the Introduction to the *ISC Security Design Criteria*, users of the *Facilities Standards* may not be aware of the context in which the security criteria are to be applied, or of how the design criteria relate to the risk guidelines and the process of drafting project-specific criteria. The applicability statement of the *Facilities Standards* is also less specific than that contained in the *ISC Security Design Criteria*. The committee views this as a critical obstacle to implementing a performance-based process because of the need to link security design solutions to higher goals and objectives for the performance of federal buildings.

Security Database

Section 5 of Executive Order 12977 requires that agencies with security responsibilities be encouraged "to share security related intelligence in a timely and cooperative manner" and that they develop and maintain "a centralized security data base of all Federal facilities." This is a necessary step for providing feedback to the threat assessment and risk analysis process and informing decision making about protective measures. The new Physical Security and Hazard Mitigation Committee of the Federal Facilities Council partially fulfills this objective, but a more formal process is warranted.

Assessing and Managing Risk

Improvements to the ISC Security Design Criteria related to risk assessment cannot be achieved without structural changes in the way the federal government manages its general approach to building security. Risk assessment requires broad-based activity by the different disciplines affected by security: physical security, architecture, civil engineering, blast effects, biological/chemical hazard effects, fire safety, and mechanical/electrical operations, especially as related to post-event life safety and rescue efforts. At the risk assessment stage, a detailed knowledge of the assessment methods used in each of these areas is not required: practical knowledge and general rules of thumb are usually sufficient. Historical cost information is also needed for analysis of alternative risk abatement techniques.

Many if not all of the required resources are available in the federal government, but interagency cooperation is essential if these resources are to be available for every project where the ISC Security Design Criteria are used. Specifically, the effort must be made to collect actual costs related to physical security after each project, or at least for a representative number of projects. If accurate cost information is not available, budgetary constraints may mean that changes to building security features instituted late in the design process are not consistent with the risk assessment conducted to guide the design. This very real influence of budgetary realities on actual provision of security features (as opposed to whether they are merely included in preliminary designs) has enormous implications for the performance of a building in an attack, as well as for the safety of the occupants and the continuance of mission-related activities.

RECOMMENDATIONS

The committee's recommendations are presented in two groups. The first are those that the committee believes can and should be implemented in the short term to help maximize comprehension and use of the important security concepts in the ISC Security Design Criteria with minimum effort. Unless this first set of recommendations is implemented as soon as possible, the ISC Security Design Criteria will probably continue to be underused and misinterpreted by users who do not have a strong background in blast or security analysis. The second set of recommendations addresses fundamental ways that the ISC Security Design Criteria could be improved over a longer term through their modification to take into account the greatly expanded body of knowledge on blast effects and other threats (chemical, biological, and radiological) that has emerged since the criteria were first developed. These recommendations should also improve government oversight of their implementation.

Group 1—Short-term Recommendations

The committee recommends that:

1. The Interagency Security Committee give immediate consideration to the changes suggested by the committee as a result of its evaluation of the ISC Security Design Criteria.
2. The *ISC Security Design Criteria* be expanded to include basic information on how blasts affect buildings and people, as a technical basis for the protective strategies incorporated in the document (for suggested content, see Chapter 3 above).
3. A preface be added to the *ISC Security Design Criteria* that clearly establishes the authority and applicability of the document by citing Executive Order 12977 and all related executive or legislative actions.
4. The *ISC Security Design Criteria* be reviewed by a professional editor familiar with building codes and standards to improve its organization and ensure consistency in format, language, and style. This should improve both the readability and the utilization of the document.
5. Information provided in Part 3 of the *ISC Security Design Criteria* document be integrated with Parts 1 and 2, so that all information pertaining to a particular design component or system (e.g., struc-

tural or mechanical) is located together, thus clarifying the document and simplifying its use.

6. The spirit and intent of the language in the introduction to the *ISC Security Design Criteria* be incorporated in any document that references the criteria.
7. The Interagency Security Committee in concert with its member agencies begin a comprehensive review of the ISC Security Design Criteria as soon as possible. This review should include creation of risk assessment and management tools as well as policy guidance for physical protection and security to guide the development of risk reduction strategies and a performance-based design process. The review should also consider expanding the criteria to (1) include critical occupancies such as child care centers, (2) broaden the range of possible threats and countermeasures to cover chemical, biological, and radiological weapons, and (3) address the impact on protective design of such issues as emergency planning, enhanced fire protection, life-cycle costing, procurement strategy, and construction security.

Group 2—Long-term Recommendations

The committee recommends that:

8. The ISC security criteria be fully integrated into the *Facilities Standards for the Public Buildings Service* and not isolated in a single chapter of that document. This will help to ensure that security design is integrated into the overall design strategy of each facility from the very start of planning.
9. The Interagency Security Committee be empowered to commit resources to implement and monitor the security design guidelines for federal buildings, as established in the ISC Security Design Criteria and as mandated by Executive Order 12977, Section 5, part (a) (2) and Section 6, part (c). Although the actual monitoring may be outsourced if there are insufficient staff within ISC member agencies, the assessment of program performance in general should be a governmental function.
10. The Interagency Security Committee be made responsible for providing leadership and vision and crafting a strategy for instituting necessary changes to the ISC Security Design Criteria. The ISC will need to communicate its vision to those whose cooperation will be needed to achieve it, and must motivate and inspire people to "energize themselves to overcome political, bureaucratic, and resource barriers."

11. The member agencies of the Interagency Security Committee employ staff with the necessary core competencies (technical, management, communications, and financial skills) to ensure that the ISC Security Design Criteria are implemented. This includes a thorough understanding of the objectives of the security criteria, the process of establishing project-specific criteria, and the consequences of failure. If aspects of implementation and monitoring are to be delegated to administrative regions, the designated agency must provide adequate training to ensure competent interpretation and implementation of the criteria and, as performance criteria permit, assessment of alternative solutions. Where in-house capabilities or resources are limited, outsourcing may be appropriate, but program objectives and implementation should remain the responsibility of the Interagency Security Committee.
12. The designated lead agency for implementation of the ISC Security Design Criteria create a security database, and other agencies, to the extent permitted by law and national security, be directed to share information as required by Executive Order 12977. This effort should include building a database of explosive events in order to define the probabilities of different explosive charge weights.
13. As experience with implementing the *ISC Security Design Criteria* increases, the document should be expanded to include examples of best practices in building design for each security level. This will give users practical demonstrations of how specific criteria can be met.

REFERENCES

Clinton, William Jefferson. 1995. Executive Order 12977, Interagency Security Committee. Washington, D.C.: The White House.

GSA (General Services Administration). 2000. Facilities Standards for the Public Buildings Service. General Services Administration. Washington, D.C.: Government Printing Office.

ISC (Interagency Security Committee). 2001. ISC Security Design Criteria for New Federal Office Buildings and Major Modernization Projects. Washington, D.C.: GSA.

NRC (National Research Council). 2000. Outsourcing Management Functions for the Acquisition of Federal Facilities. National Research Council. Washington, D.C.: National Academy Press.

Appendix

Biographies of Committee Members

Stuart L. Knoop (*Chair*) is co-founder and vice president of Oudens and Knoop Architects PC, of Chevy Chase, Maryland. He has been involved in design for security for 25 years, particularly for the U.S. State Department, Office of Overseas Buildings Operations. He has extensive experience designing security upgrades for more than 60 embassies and consulates worldwide and extensive service with the National Research Council (NRC). Mr. Knoop served on the NRC Committee on Research for the Security of Future U.S. Embassy Buildings, was vice chair of the committee that produced the NRC report *Protecting Buildings from Bomb Damage*, and was a member of the Committee for Oversight and Assessment of Blast-effects and Related Research. He is also a former member of the NRC Commission on Engineering and Technical Systems. Mr. Knoop is a registered architect, a fellow of the American Institute of Architects, and a member of the American Society for Industrial Security and the Construction Specifications Institute. He holds a B.Arch. from Carnegie Institute of Technology (Carnegie Mellon University).

Alfredo H.S. Ang, NAE, is a research professor of civil engineering at the University of California, Irvine. His areas of expertise are structural and earthquake engineering and risk and reliability analysis. He has extensive experience in applying risk-based decision criteria to structural designs for hazard mitigation and has been recognized for developing practical and effective methods of risk and reliability approaches to engineering safety-and-design structural criteria formulation. Dr. Ang is the recipient of the following awards from the American Society of Civil Engineers: Walter L. Huber Civil Engineering Research Prize, State-of-the-Art Civil Engineering Award, A.M. Freudenthal Medal Award, Senior Research Award, N.M. Newmark Medal, and the Ernest Howard Award. He is the author of numerous books and papers on risk-based design and earthquake engineering. He holds a B.S. in civil engineering and an M.S. and a Ph.D. in structural engineering.

Niall Kelly is a project architect and associate partner with the firm of Langdon Wilson Architects in Los Angeles, California. He has extensive experience in developing the technical aspects of the designs of the firm's joint venture partners on complex building projects and is familiar with applying the General Services Administration guidelines for blast-resistant construction. He served as project manager for the Lloyd D. George Federal Courthouse in Las Vegas, Nevada, which was the first U.S. courthouse to be designed and built in accordance with the GSA guidelines. Mr. Kelly is a registered architect in the state of California and a member of the American Institute of Architects. He received a bachelor's degree in architecture from the University of Southern California.

Brian Meacham is principal risk consultant and principal fire consultant with Ove Arup and Partners in their Massachusetts office. He has extensive experience in fire protection engineering as well as expertise in fire protection and particularly in the development and application of performance-based codes for life safety. Dr. Meacham has served as a consultant to private and institutional clients worldwide and directed research and technical activities at the Society of Fire Protection Engineers. He is a registered engineer in Connecticut and a chartered engineer of the Institution of Fire Engineers in the United Kingdom. He is a member of numerous technical societies, including the American Society of Safety Engineers, the Society of Fire Protection Engineers, the Institution of Fire Engineers, and the National Fire Protection Association. He holds a B.S. in electrical engineering and an M.S. in fire protection engineering from Worcester Polytechnic Institute and a Ph.D. in risk and public policy from Clark University.

Randall Nason is a corporate vice president and manager of the Security Consulting Group at C.H. Guernsey & Company in Oklahoma City, Oklahoma. An expert in physical security systems, he has more than 20 years of domestic and international experience in strategic planning, project management, facility vulnerability analysis, and the design of facility security systems, including intrusion detection, closed-circuit television, and electron entry control. Prior to joining C.H. Guernsey & Company, he was a member of the technical staff at Sandia National Laboratories, where he was responsible for investigating and resolving safeguards and security issues at nuclear facilities of the U.S. Department of Energy. Mr. Nason has published widely on topics ranging from security system project development to emerging threats and security system management. He is a registered engineer in Oklahoma and past chair of the Standing Committee on Security Architecture and Engineering of the American Society for Industrial Security. He holds B.S. and M.S. degrees in nuclear engineering from Kansas State University.

Charles Oswald is a principal engineer with Wilfred Baker Engineering in San Antonio, Texas. He has designed blast-resistant structures to withstand a wide variety of explosive threats, including close-in loading from high explosives and propellants, far-range blast loading, and leakage pressures from external explosions. His extensive experience with the performance of nonstructural materials in a blast environment includes more than 100 shock-tube tests of specialty products for windows that mitigate glass fragment hazards. Dr. Oswald is currently the program manager for a 3-year effort to develop a model to predict injuries to occupants from blast-loaded buildings. He is a registered engineer in the state of Texas and a member of the American Society of Civil Engineers and the American Concrete Institute. He holds a B.S. degree in civil engineering from Catholic University, an M.S. in civil engineering (structures) from Northwestern University, and a Ph.D. in civil engineering (structures) from the University of Texas at Austin.

Thomas Rust is chair of Patton Harris Rust & Associates. He is a registered civil engineer and certified land planner with over 35 years of diversified experience in site planning and development and is well versed in site design and construction issues. Mr. Rust has served as principal-in-charge for numerous site evaluation, development feasibility, and master planning efforts for public and private clients. He is a registered civil engineer in Virginia and has been certified by the American Institute of Certified Planners. He holds a B.S. degree in civil engineering, an M.S. in public works engineering, and a master's in urban and regional planning. He currently is the mayor of Herndon, Virginia, a position he has held for 19 years.

Kenneth Schoonover is president of KMS Associates, Inc., a codes and standards consultancy for the public and private sectors. He was formerly vice president of the Codes and Standards Department of Building Officials and Code Administrators (BOCA) International, where his responsibilities included the administration of BOCA's staff support for development of the International Code Council's (ICC) *International Codes* and production of commentaries for the *International Codes*. BOCA's participation in the development of standards promulgated by other organizations was also carried out through its Codes and Standards Department, as were the maintenance and updating of those standards that are referenced in the *International Codes*. An expert in the development and application of building codes, he is a registered engineer in Wisconsin and a member of various professional associations and societies. He holds a B.S. degree in architectural and building construction engineering from the Milwaukee School of Engineering and an M.B.A. from the Keller Graduate School of Management.

Robert Smilowitz is a principal with Weidlinger Associates of New York. He has 24 years of experience in mathematical modeling and dynamic response calculations for ship, satellite, and hardened and conventional structures subjected to shock and vibration loading. His expertise is the blast-resistant design of structures and the analysis of structures' vulnerability to vehicle-bomb attack. He has participated in the design of numerous federal courthouses and office buildings, embassy structures, airline terminals, and commercial structures. Dr. Smilowitz is a registered engineer in New York and California and has published extensively on issues in blast resistance and seismic design. He holds a B.S.C.E. from the Cooper Union for the Advancement of Science and Art and an M.S. and a Ph.D. in civil engineering from the University of Illinois at Urbana-Champaign.

James C. Snyder is a professor of architecture and urban planning and director of the Studies in Urban Security Group at the University of Michigan. He specializes in architectural and urban security as applied to crime prevention and control, physical and operational security, strategic security planning, and the security of public water systems. His expertise includes planning effective security strategies and designing (or modifying) environments to achieve desired levels of security without adversely affecting human endeavors that occur in the environment. He holds a bachelor of architecture degree from Ohio State University and a master of architecture, a master of city planning, and a Ph.D. from the University of Michigan.

Leonard C. Zimmerman is a senior vice president at Flack & Kurtz, Inc., in New York. He has over 25 years of experience in engineering, primarily in the commercial and institutional building consulting arena. His expertise is in institutional electrical systems and building systems engineering. His major accomplishments at Flack & Kurtz include the Health Sciences Center at Syracuse University, the World Bank in Washington, D.C., the Federal Office Building at Foley Square in New York, and the Naval Aviation Systems Team Complex at Patuxent River Naval Air Station. Mr.

Zimmerman is a registered engineer in 26 states and a member of the Illuminating Engineering Society, Institute of Electrical and Electronics Engineers, Association of Energy Engineers, and the National Society of Professional Engineers. He holds a B.S. degree in electrical engineering from Louisiana State University.